I0669623

Clarence Moores Weed

Seed Travellers

Studies of the Methods of dispersal of various common Seeds

Clarence Moores Weed

Seed Travellers
Studies of the Methods of dispersal of various common Seeds

ISBN/EAN: 9783337209230

Printed in Europe, USA, Canada, Australia, Japan

Cover: Foto ©berggeist007 / pixelio.de

More available books at **www.hansebooks.com**

SEED-TRAVELLERS

STUDIES OF THE METHODS OF DISPERSAL OF

VARIOUS COMMON SEEDS

BY

CLARENCE MOORES WEED

———•—

BOSTON, U.S.A.

GINN & COMPANY, PUBLISHERS

The Athenæum Press

1899.

INTRODUCTION.

THERE are few subjects better adapted to awakening the faculties of observation to a sense of the significance of those things in the living world with which we come in daily contact than that of the dispersal of seeds. Away from the crowded streets of cities one can scarcely step out of doors without witnessing some phase of plant dispersal, while a little intelligent attention to the commonest objects along the roadside will reveal numberless interesting facts.

These studies may be pursued to advantage at any season of the year, but there is an especial wealth of material during the months of autumn and winter.

For studies of nature in the schools, seeds and fruits are particularly desirable. Specimens illustrating the various methods of dispersal are easily obtained by the pupils themselves. The significance of the several adaptations to dissemination is at once apparent; the connection between the plant and its surroundings is shown to advantage ; and the general idea of the unity of nature may readily be brought before attentive minds.

I would recommend that this little book be used in connection with observations upon the specimens treated of ; that the studies be read by the individual pupils, either with the objects in hand or for the purpose of inciting them to search for the specimens. If the material does not present itself in the order of the studies as they appear in the book, the studies may be read in the order in which the specimens are obtained. It may then be advisable, after most of the parts have been read, to review the whole subject by having the pupils begin at the first of the book and read it through consecutively, with or without studying the objects again. They will thus be given a logical idea of the subject as a whole, and the knowledge already gained will be more firmly fixed in their minds.

To the teacher about to take up the subject of seed dispersal I would commend the spirit of these lines, written some years ago by Prof. L. H. Bailey : "The studious observer of nature is constantly impressed with the unlimited numbers of curious little contrivances and peculiar habits by means of which the commonest plants and animals are prepared to overcome the obstacles which surround them, for be it known that even plants have obstacles to surmount if they perpetuate their species. A plant must hold its own against its stronger and more aggressive neighbors, or suffer the fate of many of our native plants, which have been driven out by Old World weeds ; it must possess some means of scattering its seeds beyond the limits of severe competition ; it must struggle against uncongenial climate and the ruinous changes wrought by man ; and it must elude or repel the attacks of herbage-loving or seed-loving animals. One who is interested in the fascinating peculiarities of common objects is often pained by the sneering estimate put upon them by less observant people. No one is prepared to study nature so long as he regards any phenomenon, however slight in itself, as trivial and unworthy his regard. He must not attempt to play the critic with nature. He must assume the attitude of a patient learner, who accepts all things as worthy his study and consideration." [1]

C. M. W.

[1] Talks Afield about Plants.

CONTENTS.

SEED-TRAVELLERS.

PART I. — THE WIND AS A SEED DISTRIBUTER.

ON THE WINGS OF THE WIND.

EVERY one who has wandered along a country road has noticed the peculiar seed pods of the common milkweed, represented in the frontispiece. About midsummer you may see them forming as slender green cases, which increase rapidly in size. During the latter part of summer they change from green to brown, gradually drying out as the weeks go by. Finally they ripen and split open along one side.

As the pods open there is revealed a large number of flattened brown seeds, having the margins thinner than the middle portions. When the pod is wide open you may see that the seeds overlap in a manner suggestive of the shingles on a house. The ends only of most of them can be seen. Those at the tip of the pod, however, are visible throughout their length. They show that each seed bears on its smaller end a tuft of silken hairs. When the pod first opens these hairs lie straight and flat, the ends of the hairs being caught in the folds of the membranous partition which runs through the center of the pod.

On exposure to the air the folds relax their hold upon the hairs, which thus become free at their upper ends. Then each

1

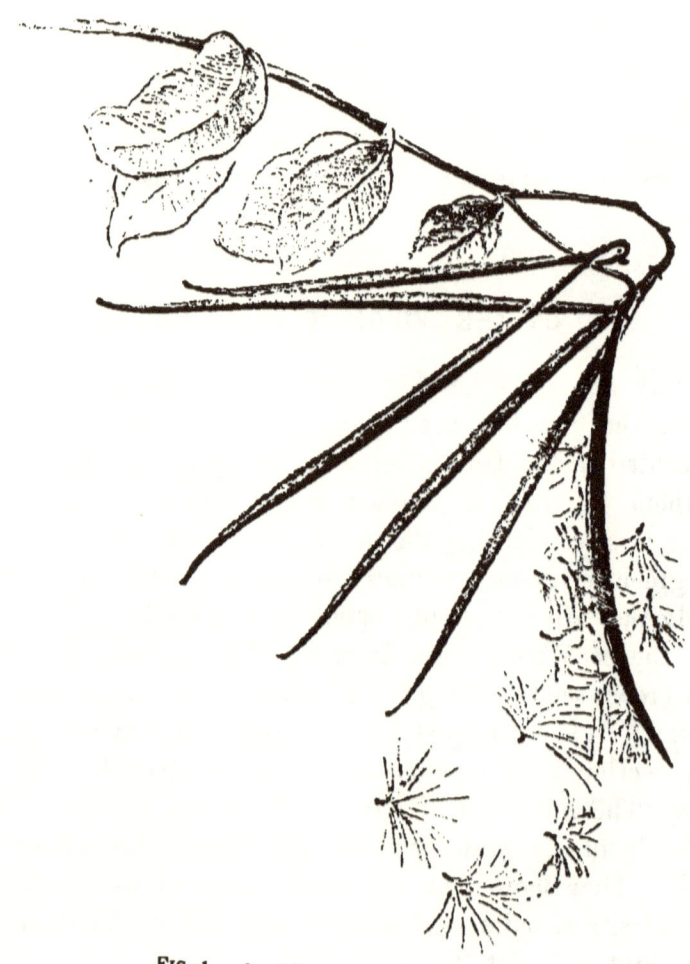

FIG. 1. — Seed Pods of Dogbane.

hair curls over toward the other end of the seed, until at last nearly all the hairs on the upper seeds are thus curled over, forming a beautiful crown, almost as light as air. When a strong wind blows, the seeds are picked up by means of these hairs and carried away to be dropped beside some fence or tree or bush.

By the beautiful device of this feathery crown the milkweed provides for the scattering of its seeds. It seems a simple process, but as you think it over you see that it is a very admirable one. It is well worth while to bring to the schoolroom two or three of the nearly ripened pods and place them beside a sunny window, or on a desk in some other part of the room. As the pods dry they crack open; the seeds become loosened; the white threads curl up. There being no wind to carry the seeds away, they form a fluffy white mass, with brown spots where the seeds show. Blow upon the mass and see how easily the tiny balloons are wafted away.

The long, slender, decorative seed pods of the spreading dogbane or Indian hemp, a plant closely related to the milkweed, give forth their beautiful little seeds in a similar way (Fig. 1).

Another example of this method of seed dispersal is found in the common fireweed or willow herb. The seeds are produced in long pods, which when mature split from above downward into four longitudinal divisions, each of which curls back and exposes the achenes to the air. Each achene is provided with a mass of tiny white hairs on its upper end. As the air reaches these they expand into parachutes, become separated, and sail away (Fig. 2).

The seeds of willow and poplar are covered with white downy silk, by means of which they are borne through the air in summer, often so filling it as to suggest a light snowstorm.

Of course, plants which rely upon the wind for the dispersal of their seeds have to take their chances that the seeds will find lodgment under conditions favorable to growth. A large majority of the seeds must be lost and never develop into plants. But so many seeds are produced that if only a small proportion are successful, the plant, as a species, will flourish.

FIG. 2. — Seed Pods of Fireweed.

The family that as a whole has most availed itself of the seed-carrying properties of the winds is the *Composite* family, — the great order of plants with flowers in heads, of which the thistle, sunflower, dandelion, and daisy are familiar examples.

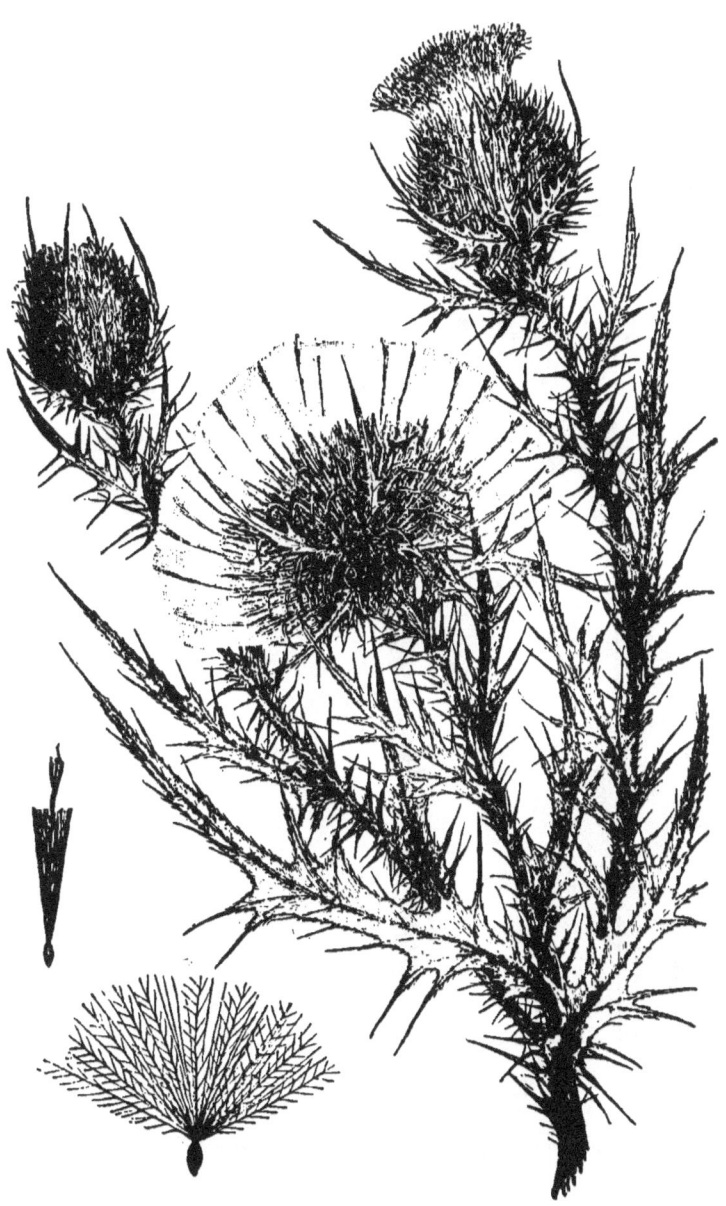

FIG. 3. — The Common Thistle. (After VASEY.)

This adaption to wind dispersal is beautifully shown in the seed-heads of the common pasture thistle. The seeds — which the botanist calls the achenes — are borne in the familiar spiny flower cups which spread apart as they ripen and dry. On

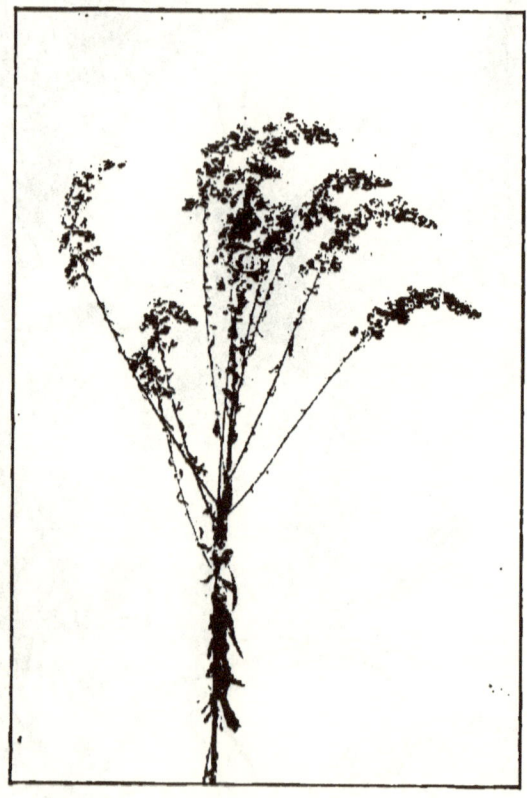

FIG. 4. — Seeds of Golden-rod.

the top of each achene is a crown of slender, white, plumose bristles, which, on exposure to the air by the spreading flower head, expand more and more, until finally the bristles are spread as a tiny parachute. Then each escapes, taking with it the seed.

This escape is most likely to take place during a drying wind, when the seeds will be carried with the breeze. They may go far before striking anything; but should they be blown against a fence or wall or stump, the seeds separate from the parachute and drop to the ground.

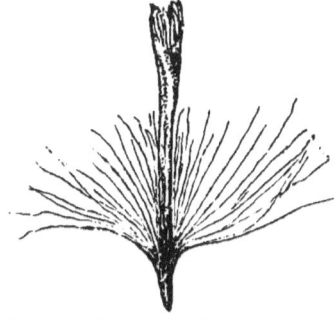

FIG. 5. — Golden-rod Seed, showing Pappus and Withered Flower.

The plumose bristles which form the parachute of the thistle seed represent the divisions of the *sepals*, which every schoolboy nowadays knows form the calyx or outer covering of the simple flowers. When the calyx is modified in this way it is usually spoken of as the pappus. The carrying power of the thistle pappus is greatly increased by the numerous plumose branches along each division.

The fact that the pappus is the modified calyx is easily seen by examining with a lens the newly ripened seed-head of an aster or golden-rod (Fig. 4). You will notice that the withered corolla of each flower, enclosing the stamens and stigma, is still in position with the limbs of the pappus surrounding it at the base (Fig. 5). At the slightest touch the corolla breaks off,

FIG. 6. — Seed-heads of Dandelion.

leaving simply the seed surmounted by the pretty expanded ring of white bristles.

The seed-heads of the aster and golden-rod are small, so that there is room for the expansion of the pappus on the rather short seeds.

The dandelion (Fig. 6) shows a slight modification of the seed structure prevailing in most compound flowers. The pappus, instead of springing directly from the top of the seed, is borne on the end of a long " beak " formed by the lengthen-

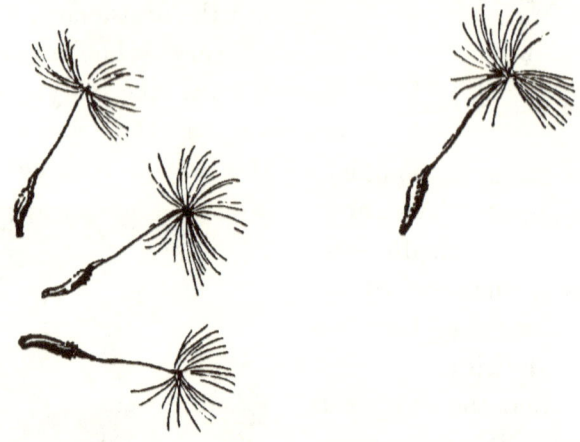

FIG. 7. — Seeds of Dandelion.

ing of the tip of the ovary. One advantage of this is to be found in the fact that by thus enlarging what we may call the circumference of expansion, the pappus of all the achenes finds room to expand. Were the pappus attached directly to the top of the ovary, there would be a very crowded condition of things when the silken tufts attempted to spread out.

After the ovules of the dandelion have been fertilized the seed-head closes up, remaining in this condition until the seed ripens. Meanwhile the beaks on the ovaries lengthen and the flower stalk grows longer, pushing the seed-head above the surrounding grass, where the ripened seeds, with the pappus, assume the form of a ghostly sphere and are picked up by the wind to be wafted far and wide.

If you examine the achene of the dandelion you will find its outer surface roughened by projecting points. By means of these, when once it lodges in the ground it is securely anchored in place.

No plant adds a more decorative effect to the scenery along the margins of lakes and ponds than that which is familiarly known as the "cat-tail" or "cat-tail flag," though sometimes

FIG. 8. — The Home of the Cat-tail Flag.

called the bulrush, and occasionally known by its botanical name, Typha.

The long, flat, straight, light-green leaves of this plant project obliquely upward above the water, on each side of the straight, smooth, cylindrical leaf-stalk, crowned with the larger cylinder of the brown seed-mass. The grace of this in turn is emphasized by the smaller, more steeple-like stem that projects above it. The whole forms a charming study in a simple decorative design, the beauty of which artists have long appreciated.

The plant is equally interesting to the botanical student. From our present point of view it is well worth while to pick

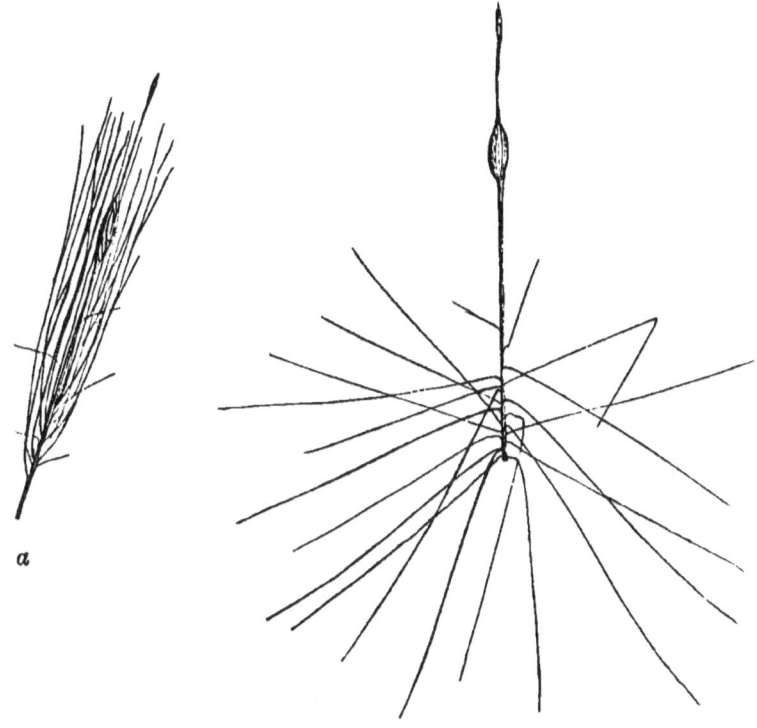

FIG. 9. — Seeds of the Cat-tail Flag.

off a little of the brown fuzz from the stalk and examine it under a simple lens. If you pull out a tiny bunch from a seed-head which has not yet begun to expand, you will see at first that the bunch consists of a great number of slender stalks, each of which has numerous small white hairs arising along its surface and lying parallel with it (Fig. 9, *a*).

But soon after you have removed the mass to a table or the stage of a simple microscope you will see it gradually become fluffy, and it will soon occupy many times the space it did at first. Look carefully now and you will see that each of the little side branches instead of lying parallel with its stem is at right angles to it. On the tiny stem above these hairs is a little oval brown body which contains the seed (Fig. 9, *b*). The stem mentioned is the stalk of the fruit, so that in this case the parachute is developed on the stalk below the ovary, instead of above as with the dandelion and other plants. Blow gently upon the fluffy mass and see how the seeds scatter.

You are now in position to appreciate better the meaning of the masses of " cat-tails " to be found by the side of nearly every pond. All through the winter the brown seed-masses project above the ice and snow, where they are visited by many seed-eating birds which peck the heads apart. Thus exposed to the air the tiny parachutes open, forming great fluffy masses that are taken up by the wind and scattered in every direction. Of course the vast majority of them will never be carried to places favorable to their growth, but a few are almost certain to reach the borders of ponds or swamps, where suitable conditions exist. If you attempt to estimate the number of seeds in a single head you will be convinced that if only one seed in ten thousand grows, the plant will be able to multiply rapidly.

In case birds do not peck at the heads, they finally are broken open by the action of the wind and weather, and the seeds are scattered in a similar way.

KEY-FRUITS, OR SAMARAS.

THERE are many methods by which seeds have been adapted to dispersal by the wind. The degree of adaptation is greatly varied. With the fruits of many trees the seed-envelopes have been drawn out into thin plates, by means of which in a strong wind — when of course they are most likely to break away from the stem — they may be carried to a considerable dis-

FIG. 10. — Maple Keys.

tance before falling to the ground. Even then during high winds many of them will be picked up and carried farther.

The familiar fruits, or "keys," of maple (Fig. 10) and ash at once come to mind as examples of this kind of dispersal. It is to be noted that generally in such cases the seed has a decided advantage in starting at a point some distance from the ground. Its chances of going far afield are much greater than they would be if the seed was borne on a herbaceous plant only a foot or two high.

Botanically speaking, the object which is commonly called the seed of maple, ash, or elm is really a fruit. While most of us think of an edible pear or apple, peach or grape, when the word fruit is brought to mind, it means to the botanist simply " the seed-bearing product " of a plant, whether edible or not.

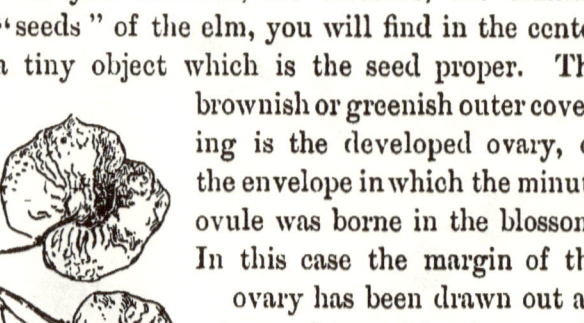

If you examine, for instance, the familiar "seeds" of the elm, you will find in the center a tiny object which is the seed proper. The brownish or greenish outer covering is the developed ovary, or the envelope in which the minute ovule was borne in the blossom. In this case the margin of the ovary has been drawn out all around into thin plates, making a tiny parachute, which in a strong wind will

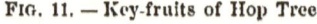

FIG. 11. — Key-fruits of Hop Tree.

sail some distance in the air before reaching the ground. Such a winged fruit is often called a samara, or key-fruit.

The hop tree, or shrubby trefoil, has a similar but larger fruit (Fig. 11) with two little black seeds in the swollen center. This is a two-celled samara, with each ovary having one-half of its margin drawn out, and the two united in such a way as to give an appearance very similar to that of the seed of the elm.

In the case of the ash, the fruit, instead of having wings all around, has a wing at the tip end only, the seeds proper being held in pockets at the basal end (Fig. 12). The fruit of the maple is a two-keyed samara, joined at the base, with the wings developed along the outer edges.

Many people have noticed that old pine cones have the plates spread wide apart, while in those recently formed they are closely appressed. If in autumn you gather one of the slender cones and hang it over a table in a warm room, a few days later you will notice that the plates are spreading open, and still later you will find on the table a number of small brown winged seeds (Fig. 13). These are the seeds of the pine. Had the cone been out of doors, the seeds, in-stead of dropping straight

FIG. 12.—Key-fruits of Ash.

downward, would have been wafted away some distance by the wind. The cones with their seeds do not mature until the second autumn after the blossom was pro-duced.

This development of wings upon the seeds or seed-cover-ing is not confined to trees and shrubs. Many herbaceous

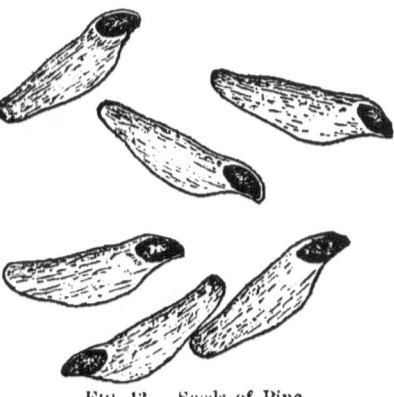

FIG. 13.—Seeds of Pine.

FIG. 14. — Leaf and Fruits of Yellow Dock. (After VASEY.)

plants have also adopted it, although in such cases the fruits are generally smaller than those of trees. A familiar example is found in the common yellow dock, represented in Fig. 14. If you examine the fruits this plant produces so profusely in summer and autumn, and to be found more or less abundantly through the winter months, you will see that the seed-covering is developed into thin plate-like margins, which greatly increase the surface exposed to the wind.

FLY-AWAY GRASS AND TUMBLE-WEEDS.

ONE breezy October morning the neighboring fields presented the appearance of a fairies' carnival. A thousand tenuous will-o'-the-wisps were dancing and sailing and whirling in every direction. Now one alone with feathery grace

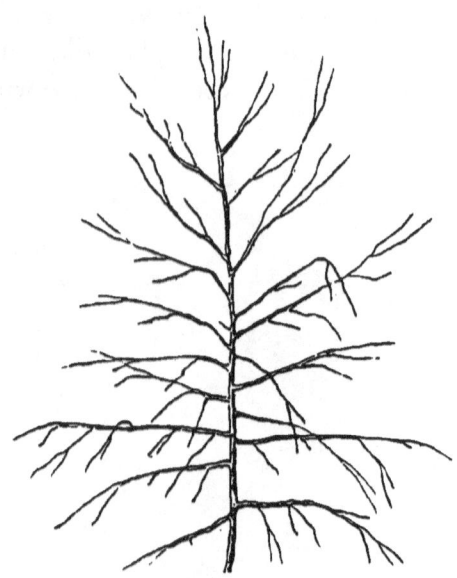

FIG. 15. — Head of Fly-away Grass.

would glide along, to join a moment later a host of airy sprites, and be wafted hither and thither by the erratic breath of the zephyr god. Here and there the paths of miniature cyclones could be traced by the movements of whirling circles, while in other places solid phalanxes moved steadily forward. The

ranks of the revellers were constantly depleted through deser-
tions to the eastward, to be quickly filled by new recruits
from out the west.

With some difficulty I caught a few of these feathery sprites,
and, holding them securely, started homeward. But a sudden
gust of wind left me empty-handed, save for some. tiny pieces
of stems; the sprites, again at liberty, sailed away with mocking
grace. I caught more, and, shielding them from the wind, got
them safely indoors, where they proved to be the seed-heads of
a grass commonly known as the " old-witch grass."

The seeds of this plant are produced in a long, wide-spread-
ing head (Fig. 15) called by botanists a
panicle. The lower branches of the panicle
curve downward, while the upper ones
curve upward, thus giving to the panicle
as a whole a rounded outline well adapted
to rolling along the ground. The branches
are joined to the stalk by thickened braces
(Fig. 16), making the union much firmer
than it otherwise would be.

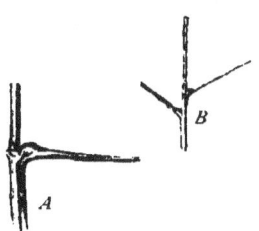

FIG. 16.—Joints of Old-witch
Grass, showing braces :
A, lower branch ;
B, upper branches.

The stem below the panicle is very
brittle. As soon as the seeds are thoroughly ripened the stem
becomes dry and is broken off by the wind. The seed-head is
then wafted away until stopped by some obstacle. The seeds,
held in tiny pockets at the tip of the branches, drop out on the
way, so that the panicle scatters them all along its path.

The old-witch grass has thus adopted a most efficient method
of seed distribution. Out of the hundreds of seeds sown broad-
cast by every whirling panicle, some are pretty sure to find
the right conditions for growth.

This old-witch grass, or "fool hay" as it is sometimes called,
may serve to illustrate the method adopted by a large number

of plants for the dispersal of their seeds. Other grasses, notably the fly-away grass, have taken advantage of it, as well as many of our most noxious weeds.

The various tumble-weeds derive their common name from the habit of tumbling or rolling along the ground when the wind is blowing, scattering far and wide their myriad seeds. These plants usually have an oval or spherical outline, and the stem breaks off above the root after the ripening of the seed. One of the most familiar examples is the common tumble-weed of waste grounds, — the *Amaranthus albus* of botanists.

In the great, unbroken sweep of the prairies the tumble-weeds are especially at home. In such regions they flourish much more than in hilly or mountainous localities, because of the comparatively few obstacles to prevent their wide dispersal.

THE RUSSIAN TUMBLE-WEED.

ABOUT twenty years ago a colony of immigrants brought from the plains of southern Russia to the prairie region of Dakota a small quantity of flaxseed.

The flaxseed was sown in the fertile soil of the new home. It sprouted and grew. Along with it there also developed a slender, reddish plant which seemed natural enough to the immigrants, for it had been commonly present in the crops on the far-away prairies from whence they came.

The slender, reddish plants waxed strong and as they grew older broadened out, becoming harsh and spiny. When the flax was harvested the spiny plants were probably left in the field. They were not useful to the flax crop, and in the density of his ignorance one could not expect the immigrant to see in those scattered plants a menace of tremendous import to American agriculture. The eyes of others were equally blind. The following season more plants came up, and so the species continued to multiply year after year.

This plant first appeared in a locality which was wooded and hilly, but in a few seasons it reached the adjacent plains, where it was rolled by the wind for miles and miles, each year afterward invading new territory. Within a dozen years it had spread throughout South Dakota, had entered North Dakota on the south, Iowa on the north, and Nebraska on the east. During the next few years it spread with marvellous rapidity, invading Minnesota, Wisconsin, Colorado, Illinois, and Ohio. Its progress was aided by the railroads that carried the seed to many distant localities, which quickly became new centers of distribution. Presumably the plant will continue to spread by similar methods, and within a few years will be present in most of the United States.

Such is the past history, so far as it can be traced, of the plant commonly called Russian thistle or Russian cactus (Fig. 17), although it is neither a thistle nor a cactus. More appro- priately it is sometimes spoken of as the Russian tumble-weed. Botanically, it is a saltwort, being considered merely a variety

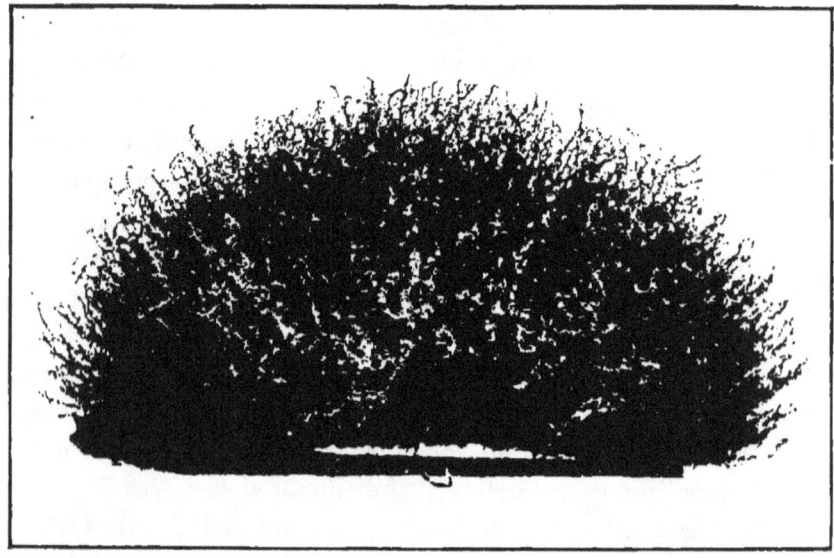

FIG. 17. — The Russian Thistle. (From SELBY.)

of the saltwort common along our Atlantic coast, as well as in many parts of Europe.

In the plains region of southeastern Russia this plant has long been known as a noxious pest. On its account "the cul- tivation of crops has been abandoned over large areas in some of the provinces near the Caspian Sea." In our own west it has already caused damage amounting in a single state to mil- lions of dollars a year.

The Russian thistle begins its yearly growth in a simple, inoffensive way. The young plants are slender and succulent, but as they grow older they harden and spread out, becoming densely covered with sharp spines. When full grown they often reach a diameter of four or five feet, a majority of the specimens being distinctly rounded in outline. After the

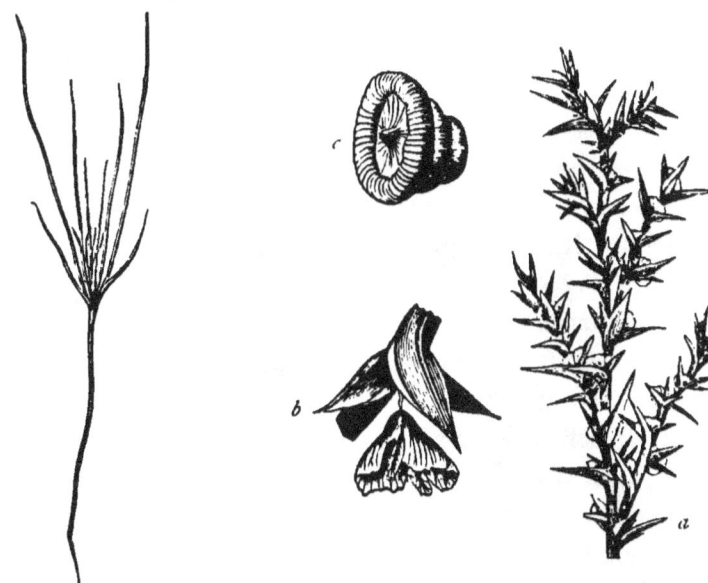

FIG. 18. — Young Russian Tumble-weed. (After DEWEY.)

FIG. 19. — Russian Tumble-weed: *a*, part of branch, natural size ; *b*, flower held by threads as in the rolling plant, magnified ; *c*, seed, magnified. (After DEWEY.)

seeds have matured the stem twists around and breaks off, thus leaving the plant to roll wherever the wind blows it, dropping its seeds as it goes along. As one large plant sometimes produces 200,000 seeds, and may be blown for miles, one can readily imagine how soon a prairie region might be overrun by the pest, which grows so vigorously that it crowds out practically all plants with which it comes in competition.

SLIDING ON THE SNOW.

For a long period of each year in our northern regions the earth is covered with a mantle of snow which often becomes coated with a crust of ice. When this happens there is produced a smooth, slippery surface across which small objects readily slide before the wintry winds.

This snowy crust is an important aid in the dispersal of many seeds. The seed-bearing branches of many of our commonest plants project above the snow, where they are visited by winter birds that come to feed upon the seeds. But the birds scatter nearly as many as they devour. When the snowy surface holds the seeds from falling, the latter are in position to be driven over the surface by the wind. If there is an icy crust they are likely to go alone; if not, they may be carried with the drifting snow.

While this method of seed dispersal is universal throughout our northern states, it operates most freely in the plains regions of the west and northwest. To determine how effective this means of dispersal is, Prof. H. L. Bolley of North Dakota recently performed the following experiment. On Jan. 31, 1895, when there was a light snowfall upon crusted snow, with the wind constant from the northwest at the rate of twenty miles an hour, a peck of mixed seed was poured upon the crust. Thirty rods distant, at right angles to the course of drifting, a three-inch trench in the snow, four rods long, served to catch the drifting seeds. At the end of ten minutes the trench was found to contain:

Millet seed	Very many seeds.
Wheat	191 "
Flax	53 "
Buckwheat	43 "
Ragweed	91 "

Similar experiments were made at other times, and the conclusion was reached that " weed seeds of almost any size, as French-weed, Russian cactus, or ragweed, may travel with the drifting snows many miles during heavy storms, settle down into the snow, and there be buried in the soil upon the melting of the snow."

You may often observe a similar process of seed distribution on ponds in winter. On the side of the pond away from the direction of the prevailing wind there will be found windrows of seeds of sedges and other plants that have been blown upon the ice across the pond.

THE problems of existence which plants have to solve are innumerable. Wherever they grow they must so adapt themselves to surrounding conditions that they shall derive the greatest benefit with the least expenditure of vital force.

Different situations require different adaptations. Those plants which have adopted as their home the quiet borders of the stagnant pool or the inland lake, the reedy marsh or the slow-running creek, have to fulfill the conditions of existence by methods very different from those employed by the plants of the hillside and wayside. In the problem of the latter, as far as seed dispersal is concerned, the air, the bird, or the beast are the factors chiefly to be considered, but the solution of the problem of the former depends largely upon adaptation to the surrounding water.

The ways in which the seeds of water plants are adapted to dispersal may well be illustrated by three groups found everywhere in the vicinity of standing water, — the sedges, the arrow-leaf plants, and the water lilies.

The most abundant plants in marshes and by pondsides are the sedges. They resemble coarse grasses, for which they are frequently mistaken. Some of them have seeds adapted to wind dispersal by means of cottony tufts of hairs; but most of them simply cast their seeds upon the quiet waters, where they float upon the surface and are driven along by every breath of wind.

It will be worth your while to remove some "seed" of sedge from a ripened head and study its structure. As you

pick up what appears to be the seed (Fig. 20) you notice how
little weight it has. On looking closer you are likely to see
that it is triangular, in many species being shaped like a miniature beechnut. If you press upon it the "seed" breaks, and
you find it apparently hollow on the inside. But if you look
carefully you will see within a tiny body which is really the
seed. The other is simply an air-filled boat in which the seed
remains.

A seed with such an outer covering is called an *achene*,
although in most achenes there is not the air space which these
sedges show.

Now drop some of these sedge achenes upon the surface of
water in a tumbler or other vessel.
Do they sink? See them rest buoy-
antly upon the top, with one flat side
down and the two other sides projecting upward. Blow gently across
the water; see how quickly the tiny
sails catch the breeze and the achenes
move away. Fancy them upon a

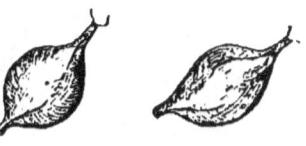

FIG. 20. — Sedge Achenes.

quiet pool out of doors: the wind ripples the surface and away
they go to the other side, where they may find lodgment, or,
perchance, if the pool has an outlet, they may be carried far
away by the running water. Either contingency fits their
needs: if the water is quiet they ride upon the surface blown
by the wind; if it is moving they are carried with it. It
would be difficult to imagine more perfect adaptation to surrounding conditions or simpler means of attaining it.

The forms assumed by the leaves of various water plants are
interesting because they vary so greatly in different kinds of
plants. Few leaves are more striking in appearance than those
of the arrow-leaf which grows so abundantly along the borders

of slow-running streams, as well as along the margins of ponds and marshes. The blossoms of the larger species are white and conspicuous, being borne on good-sized stalks that project above the surface of the water. The fruits are developed on these stalks, and the seeds are surrounded by thick, air-filled tissues, so that they float readily upon the water.

The fruits of our two common kinds of pond lilies are also adapted to sailing on the water. In the case of the white water lily — which the botanist places in the genus *Nymphæa* — " each seed is enveloped in a coat which loosely clothes the

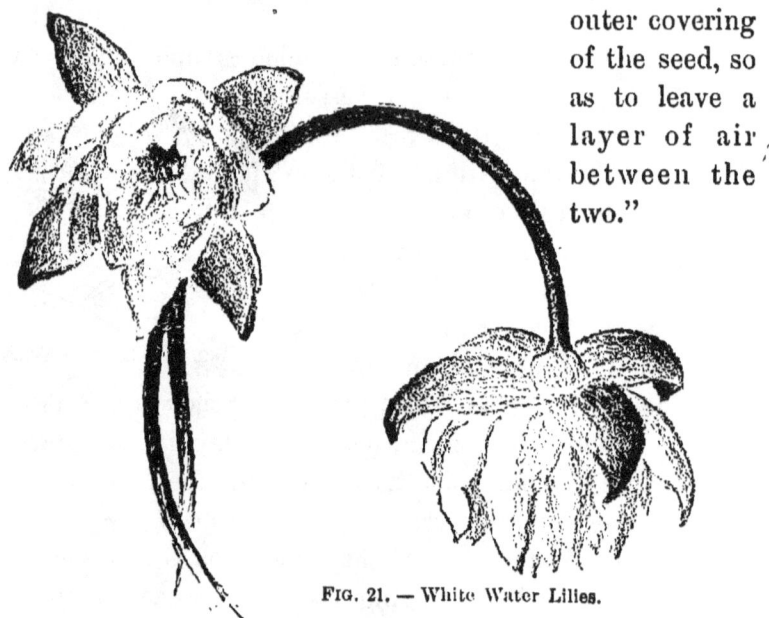

outer covering of the seed, so as to leave a layer of air between the two."

FIG. 21. — White Water Lilies.

In our yellow pond lilies — belonging to the genus *Nuphar* — there are two layers to the ripe fruit, " of which the outer one is green and succulent, while the inner one is white and charged with air, and encloses a large number of seeds." The fruits of both float upon the water, being driven over the surface by the wind or carried by the motion of the currents.

PART II. — SEED DISSEMINATION BY BIRDS.

TRAVELLING WITH THE BIRDS.

IN the struggle for existence among plants and animals, advantage is taken of all sorts of conditions that may aid in

FIG. 22. — Bittern among Sedges.

the multiplication of the species. If you put your fingers into a tumbler of water having the seed-like achenes of a sedge floating on the surface, and then withdraw them, you are likely to find one or more of the achenes attached to your fingers. When a wild duck swims around upon an inland pond where these sedge fruits are floating, it comes in contact with many of them ; when it rises to fly to another pond, perhaps far distant, some of the fruits are very likely to adhere to the feathers, remaining in position until the duck settles in a pond or lake once more, when they will again float upon still waters new to them. In this way, especially during the spring and autumn migrations of waterfowl, these seeds are likely to be dispersed over wide areas, and thus be constantly introduced into new localities.

These sedge achenes must finally be broken up, and many of the tiny seeds within the husk will settle in the mud along the borders of the pond. But even then, the possibilities of their dispersal are by no means exhausted. By the pondside live the herons and cranes, the snipes and sandpipers, the rails, plovers, and coots, and in the south the flamingoes and pelicans. Here also come, especially in early summer, the swallows and martins, the robins and thrushes, as well as other flying birds that seek the water to bathe or to drink. Nearly all of these birds, especially the larger ones, wade in the mud, and when they fly more or less of it must adhere to their feet and be carried to new localities, to be mixed with the mud of other pondsides. Thus seed distribution of sedges and other water plants must take place to a great extent through these birds, many of which are known to fly rapidly over long distances.

This seems a very simple matter now as one reads of it, but how few of us have thought of it as we saw a heron rise slowly from the margin of the lake. Like a thousand other things in nature, it remained unnoticed

" Till one who sees came passing by."

To Darwin, the great English naturalist, whose genius consisted largely in his ability to see the significance of the little things in life, we are indebted for the elucidation of this method by which seeds travel with the birds. Many years ago he reported the following simple experiment, which may easily be repeated in any schoolroom : "I do not believe," he writes, "that botanists are aware how charged the mud of ponds is with seeds. I have tried several little experiments, but will here give only the most striking case. I took in February three different tablespoonfuls of mud from three different places, beneath water, on the edge of a little pond. This mud when dry weighed only six and three-fourths ounces. I kept it covered in my study for six months, pulling up and counting each plant as it grew. The plants were of many kinds, and were altogether five hundred and thirty-seven in number; and yet the viscid mud was all contained in a breakfast cup. Considering these facts, I think it would be an inexplicable circumstance if water birds did not transport the seeds of the same fresh-water plants to unstocked ponds and streams, situated at very distant points." Additional evidence concerning this method of dispersal has been given by the German naturalist Kerner, who examined the mud from "the beaks, feet, and feathers of swallows, snipe, wagtails, and jackdaws," and who gives a list of twenty-one species of pondside plants whose seeds he found in this mud. "Most of these species," Kerner writes, "are distributed over all parts of the world, but they seldom remain for a long time in any particular locality. They often start up quite unexpectedly at places where migrating birds have rested and gone to drink. The extraordinary occurrence on the edges of ponds in southern Bohemia of the tiny grass *Coleanthus subtilis*, which is indigenous to India, and the sudden appearance of the same species in the west of France about twenty years ago may unhesitatingly be attributed to this mode of dispersion, as may also the occurrence of the tropical sedge *Scirpus atropurpureus* on the shores of the Lake of Geneva."

But the plants of the water side are not the only ones that travel with the birds. After a rain, the muddy condition of the pond-shore is repeated over a large part of the soil surface. The earth, with myriad seeds mixed with it, is sticky, and adheres to the feet of the many birds that light upon it, as well as to those of the mammals which tread over it, and thus the seeds are carried hither and thither in every direction. Many of our commonest weedy plants having small seeds are distributed in this manner.

WILD CHERRIES.

EVERY boy who has lived in or visited the country in September knows the appearance and flavor of the wild black cherry. This tree is distributed over a large portion of the United States, and is found most abundantly along the roadside fences. In size it varies from a shrub to a tall tree, with the bark smooth and shiny, colored brown, mottled with more or less gray. The leaves are oval, smooth, and shiny, light green in color, and have finely toothed margins.

In May the beautiful racemes of white blossoms appear all along the smaller twigs. Each raceme is a tiny branch set apart for fruit production. It bears thirty or forty of the small white blossoms, which soon pass by, to be replaced in part by the newly formed cherries. These are small, green, and round, each being borne on a short individual stem attached to the main stem of the raceme. There are usually on each cluster only about a third as many fruits as there were blossoms.

As the summer days go by, the cherries slowly increase in size, retaining the green color which makes them inconspicuous among the green leaves. Late in summer or early in autumn each cherry becomes full grown, having a diameter of a little over a quarter of an inch. It now gradually changes color from green to black. When fully ripe it is a brilliant, shining black, and the racemes of fruit show plainly by contrast with the light green foliage.

Each ripe cherry has a thin skin covering the juicy pulp that surrounds the large seed in the center. The seed has a hard,

33

shell-like covering, within which is the white seed material
that contains the embryo plant.

These cherries are eaten by many birds and are largely dis-
seminated by them. Even so large a bird as the crow feeds
freely upon them, while various members of the thrush family,
such as the robin, the brown thrush, and the catbird, make of

FIG. 23. — The Common Crow. (After BARROWS.)

the cherries one of the most important parts of their food during the early autumn.

The cherries are taken into the crops of these birds. The
edible pulp that surrounds the pit is digested, while the hard
seeds are either thrown up through the mouth or pass through
the alimentary canal. In either case they fall to the ground,
frequently after having been carried some distance from the
tree on which they grew, and in due time many of them ger-
minate and grow into trees.

THE POKEBERRY, OR POKEWEED.

In many places throughout the northern states, especially in recent clearings in woodlands, the pokeberry, pokeweed,

Fig. 24. — The Pokeberry.

or garget, is one of the most familiar autumn plants. Growing generally in masses, to a height of four or five feet, the plant is conspicuous on account of its purplish red stems, light yellowish green leaves, and long bunches of brilliant, black purple berries (Fig. 24).

These berries are arranged all around the central stem, there being about fifty berries in each mass. The ripest and blackest fruits are at the base of the stem, those at the tip ripening last. In September a single fruit-bunch will show the various shades from green through purple to black, which characterize the ripening process.

Each berry suggests by its shape a miniature apple somewhat

FIG. 25 — The Mocking Bird. (After BEAL.)

flattened on the ends. The remnants of the pistils project as a circle of small pointed spurs on the top. If you break open the skin, you find on the inside a juicy purple pulp, in which are imbedded nearly a dozen small, black, flattened seeds. Inside the black seed-coat of each is the white material surrounding the tiny embryo plant.

The pokeberries remain upon the stems a long time after the frost has chilled the leaves. The berries are eaten by cedar birds, crows, blackbirds, robins, and many other members of the feathered tribes. In the southern states the famous mocking bird feeds upon them, as well as upon the seeds and berries of many other plants, such as poison ivy, Virginia creeper, sumach, red cedar, black alder, and bayberry. The German naturalist Kerner reports that "a song thrush sickened after eating berries" of the pokeweed, but there seems to be little evidence

FIG. 26. — The Cedar Bird.

to indicate that the berries are injurious to the wild birds that feed upon them here. The seeds are so small and hard that they probably pass through the alimentary canals of the birds without being digested.

The situations in which the pokeweed is most abundant — along roadsides and in clearings in woods — are suggestive of dissemination by birds, for these are the places where one is most likely to find the fruit-eating songsters.

THE barberry is another fruit that is attractive to birds. All along the coast region of New England, and in many other localities inward, this handsome shrub is found in abundance,

FIG. 27. — Branches of Barberries.

while all over the country it is commonly planted in parks and grounds for its ornamental value. It is beautiful throughout the year, especially in early summer when the graceful racemes of light yellow flowers form a pleasing color harmony with the blue green leaves, and in autumn when the drooping clusters of brilliant red berries add a unique charm to the landscape.

If you break open the skin of one of the barberry fruits, you will find inside a red pulp that surrounds a brown seed, some-what the shape of an apple seed, although it is not flattened so much upon the sides. These berries are eaten by many different kinds of birds, which scatter the seeds in every direction. In the case of most of the smaller birds that feed upon the berries, the seeds are probably ejected through the mouth after the pulp has been digested. The fact that the barberries hang upon the bushes from autumn until spring, always ready to be eaten by any feathered vagrant, even when the snow covers other kinds of food, is of great advantage to the plant.

THE POISON IVY AND THE HARMLESS SUMACHS.

THE poison ivy is everywhere abundant, and is an interesting example of a plant whose seeds are dispersed by birds. The
pretty, compound, three-parted leaves of this plant are only too
familiar to many people susceptible to the subtle poison that
renders contact with the vines a matter of serious concern.
Others, however, can handle the leaves and fruits with im-
punity.

As autumn weaves the brilliant web of her showy garment,
she uses the poison ivy and the Virginia creeper to fill in large
masses of deep crimson tones, that form a pleasing color harmony with the yellows and browns of the low-lying herbage.

More or less hidden by the foliage of the ivy are the white
berries which form its fruit. They are borne in broken racemes
on slender stems. Each berry is globular, though more or less
irregular. As you break it open you find on the outside the
brittle white skin, inside of which is a whitish firm substance,
closely connected with the seed-coat of the fruit. This seems as
unpromising a "berry" as one well could imagine, yet on
account of the dryness of what stands in place of the pulp of
other berries, it will keep through the winter in as good condition as when it first matures.

This is its method of adapting itself to the conditions of
life. During late summer and throughout autumn the birds
are able to get raspberries, blackberries, grapes, wild cherries,
and other succulent fruits. These pass, however, with the
season, and during the period when Mother Earth is shrouded
in a snowy mantle, the birds fall back upon these better-keep-

ing fruits for sustenance. From October to February the crows, for example, feed freely upon the berries of poison ivy, there being one case on record in which one hundred and fifty-three of these seeds were taken from the stomach of a single crow. Soon after the seeds are swallowed the outer part is

Fig. 28. — The Red Sumach Berries.

removed by the action of the stomach, and the seeds are thrown out through the mouth of the bird, in "pellets" similar to the one represented in Fig. 30.

Botanically, the poison ivy is a sumach belonging to the genus *Rhus*, that includes the common harmless sumachs, the most abundant of which is represented in Fig. 28. This, too, is largely used in autumn's gaily colored woof: here and there it lies in solid patches of crimson and maroon, — glowing colors that add much beauty to hillside and wayside.

These harmless sumachs are produced in large panicles on the ends of the branches. When mature they are of a deep maroon color. As one examines them individually, they seem as food supplies even less promising than the berries of the poison ivy. The birds, apparently, are also of this opinion, for in general they leave these berries until other sources of food

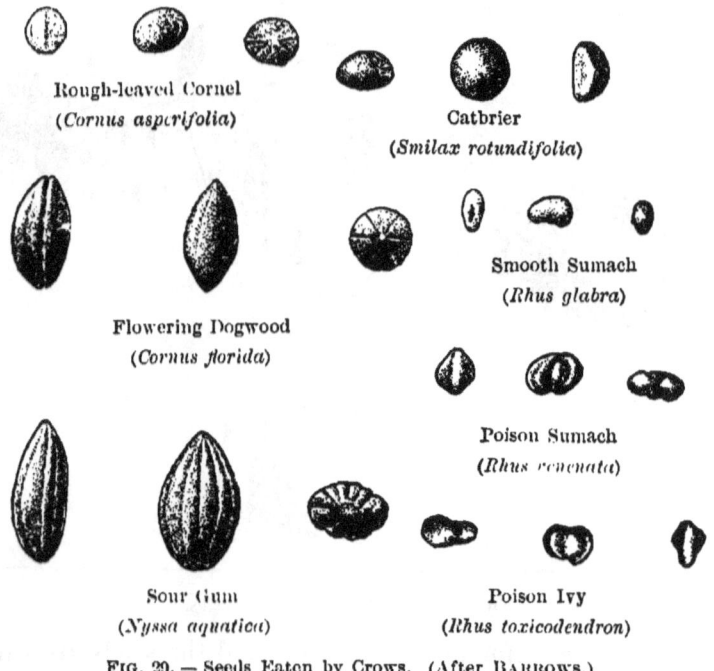

Rough-leaved Cornel
(*Cornus asperifolia*)

Catbrier
(*Smilax rotundifolia*)

Flowering Dogwood
(*Cornus florida*)

Smooth Sumach
(*Rhus glabra*)

Poison Sumach
(*Rhus venenata*)

Sour Gum
(*Nyssa aquatica*)

Poison Ivy
(*Rhus toxicodendron*)

FIG. 29. — Seeds Eaten by Crows. (After Barrows.)

are exhausted. Even the crows wait until winter before they feed freely upon them. A number of birds, the catbird, for example, eke out the slender diet of early spring with these berries.

A few years ago Prof. Walter B. Barrows made a careful study of the food of the crow. He discovered many interesting facts in connection with the life history and feeding habits of these birds, particularly with reference to the crow "roosts." These are places where immense numbers of crows congregate every night. Regarding the dispersal of seed in the vicinity of these roosts, Professor Barrows writes:

"The following facts serve to show how extensive is this seed-planting in the vicinity of roosts: On Feb. 8, 1889, I visited the well-known — almost historical — crow roost located on the Virginia side of the Potomac River, just opposite Washington, D. C. The exact location of this roost varies from time to time, but at the date mentioned it was entirely within the grounds of the National Cemetery at Arlington, and covered an area of twelve or fifteen acres of second-growth deciduous trees. The ground beneath these trees was pretty evenly covered with the ejects of the crows, forming a deposit which in places was an inch or more thick, though the average deposit was probably less than half an inch. A representative spot, free from underbrush, was selected, and all the material above the leaves from an area two feet square was carefully collected, dried, and examined. The weight of this material when dry was almost exactly one pound, and it contained the following seeds:

	NUMBER.
Poison ivy (*Rhus toxicodendron*)	1041
Poison sumach (*Rhus venenata*)	341
Other sumachs (*Rhus*)	3271
Juniper, or red cedar (*Juniperus virginiana*) .	95
Flowering dogwood (*Cornus florida*) .	10
Sour gum (*Nyssa aquatica*)	6
Total .	4764

"A little calculation shows that the roost of fifteen acres must have contained upward of 778,000,000 seeds, or more than enough to plant 1150 acres as thickly as wheat is sown.

"Of course the seeds thus dropped at the roost are subject
to such unfavorable conditions that comparatively few grow,
but it must be remembered that crows spend only the hours of
darkness at the roosts, while during at least twelve hours each
day they are dispersed far and wide over the surrounding
country, collecting and distributing these seeds. The process
of digestion — at least the preliminary process — is very rapid
in crows. A caged crow, experimented on during several

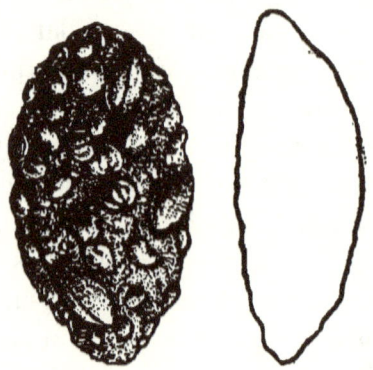

FIG. 30. — Crow Pellet. (After BARROWS.)

months in the winter of 1889–90, ate berries of poison ivy
with greater relish than any other wild fruit obtainable. He
swallowed about eighty berries within a few moments, taking
several mouthfuls of sand immediately afterwards; and about
thirty minutes later he disgorged a large pellet, consisting
entirely of sand and the seeds of the poison ivy berries, the
latter with every shred of pulp removed by the gizzard-like
action of the stomach."

THE BURR MARIGOLD AND THE HOOKED CROWFOOT.

THE burr marigold is one of the most characteristic autumn plants. Along the roadsides, in the ditches, or by the borders of the pond one is quite certain to find the large bright yellow blossoms crowning the double-branching stems. At the base of each flower head there is a double circle of large green bracts, the outer being larger and more conspicuous than the inner. Like the dandelion, the thistle, and the sunflower, the burr marigold belongs to the family of composite plants, in which many tiny flowers are crowded together into a single head, so that what we commonly think of as the blossom really consists of many individual flowers.

Inside the ring of green bracts there is a row of yellow petal-like objects, which give the blossom its chief attractiveness; remove these and the flower head becomes inconspicuous. These are the so-called ray flowers; the remainder are the disk flowers. Pull the blossom apart and you will see that the individual disk flowers have a general cylindrical appearance, with the seed-bearing part — called the ovary — at the base. In some of the older blossoms which are becoming brown, notice that these ovaries have developed into what we commonly call the seed.

Look at one of these seeds through a simple lens, and study its structure. See the four ribs extending up and down along the sides, and notice particularly the sharp-pointed hooks curv-

Fig. 31. — Pitchforks, or Beggar Ticks. (After COVILLE.)

ing backward toward the base. See how these ribs project up beyond the seed, as spines provided with recurved barbs.

In pulling the seed-head to pieces, some of these seeds are likely to adhere to the fingers by means of these barbs, while if you touch them to a piece of cloth they will "stick tight," — a fact which has given them this term for a common name. It is easy to see how this sort of an adaptation would be useful to the plant in getting its seed dispersed.

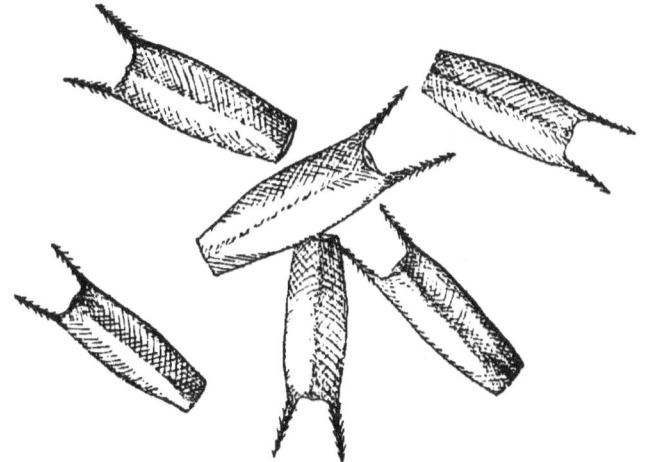

FIG. 32. — Achenes of Beggar Ticks, Magnified.

Instead of calling upon the wind to waft its seeds far and wide, it makes the beasts of the field its burden-bearers. These "stick-tights" will take firm hold upon the hair or fur of almost any of the larger animals, many of which under the conditions existing in previous ages of the world, when our plants were developing, roamed about in just the situations where the burr marigold is most at home. So, also, they do to-day, though mankind has interfered in the older settled regions to render communication by such animals between regions far apart more difficult than formerly.

Besides the common burr marigold there are several other species of plants belonging to the same genus that have adopted similar methods of seed dispersal. Perhaps the most abundant of these are the common "beggar ticks," or "pitchforks," in which the achenes (Fig. 32) have two pointed projections, called awns, and the Spanish needles, in which the achenes are larger and have three or four awns.

FIG. 33. — Hooked Achenes of Crowfoot, Magnified.

Along the borders of woods one can commonly find in autumn the peculiar seed-heads of the hooked crowfoot. The individual achenes of this plant are represented, enlarged, in Fig. 33. In each the seed-coat is prolonged into a short stalk that ends in a recurved hook. These achenes are crowded together into a small round head, and readily take hold upon the hair of any animal that may brush against them.

THE BURDOCK AND THE CLOTBURR.

IN the burr marigold each achene is provided with hooks to assist in its dissemination, but in the case of the common burdock the hooks are on the outside of the seed-head, while the seeds themselves are quite smooth.

If you examine a burdock blossom you will find the lower part of the flower head covered with green scales, each of which projects upward and outward, and at the tip curves over into a sharp-pointed hook, much the shape of a fishhook. As the flower matures these hooks gradually become dry. Finally, when the seeds are ripe, the hooks are ready to catch hold of any animal that brushes against the plant.

By this time the connection with the stem at the base of the flower head has become sufficiently loosened so that the burr pulls off readily. Yet it holds on tight enough to remain attached to the plant through the winter, unless the grappling hooks are taken hold of by some external agency. Consequently, the period during which the seeds are open to dissemination extends over many months. This, of course, is a decided advantage, for it greatly increases the chances that the seeds will be carried to other localities.

When the burr becomes attached to the hair of an animal, it may be some time before it is removed. As it is rubbed by the creature or is brushed against trees or branches, it is likely to be pushed open, and the dozen or more seeds are likely, one by one, to drop to the ground. The individual seeds are rather large, in color brown mottled with black, and rather smooth except for a few slightly projecting, longitudinal ridges.

49

FIG. 34. — The Common Burdock.

The seed-heads of the common clotburr (Fig. 36) of waste grounds are very similar in general structure to those of the burdock, as will be seen by referring to Fig. 34. The burrs are oblong, about an inch long, with two large spiny hooks at the tip, and many smaller hooks scattered over the surface. They get into the wool of sheep, whence they are not easily dislodged, so that when the wool is cut off and shipped to distant regions, these burrs are often carried with it.

An idea of the important part played by these various hooks and spines in the dissemination of seeds may be gained by reading the following paragraph written by the German botanist Kerner:

FIG. 35. — Seeds of Burdock.

"About ten per cent of all the flowering plants possess fruits and seeds which are dispersed by means of clawed or barbed processes. The part of the plant which is provided with these structures hooks on to the hairs, bristles, or feathers of any bird or other animal that happens to come into contact with it. The consequence is that it is torn away and carried off by the animal. This act of depredation is, of course, not intentional on the part of the creature that performs it: on the contrary, such appendages are a source of discomfort, and are got rid of as soon as possible. But in many cases this is not accomplished until a considerable distance has been traversed,

FIG. 36. — Clotburr : 1, Plant in Blossom ; 2, Burrs. (After VASEY.)

and sometimes the troublesome objects remain for weeks in the creature's coat or mane. The organs of attachment are either hooked at the tip or beset with barbs. In the latter case, the barbs are borne on special rigid bristles or needles, and are either collected together at the top as in a harpoon or else are arranged in longitudinal rows as in a hackle for combing flax. Only in a few instances do these structures, which may be classed together as hooked bristles and hooked prickles, occur on the seeds themselves; usually they are appendages of the seed-coat, and as such exhibit every degree of size possible, from the delicate hooked bristles on the small nutlets of the enchanter's nightshade· to the thick, firm claws of the African harpoon fruit. The hooked spines of the latter fruits attain the size of crows' feet, and are a notorious source of vexation to ruminant animals, both wild and tame." [1]

[1] Natural History of Plants, II, 871.